For many years people dreamed of a bridge that could cross the five miles of often wild and treacherous waters of the Straits of Mackinac. In November of 1957 that dream was realized when the newly completed Mackinac Bridge was opened for traffic connecting Michigan's two great peninsulas.

Sweeping panoramas of shoreline follow the margins of the great water expanse of Lake Michigan and Lake Huron and gradually converge where the lakes join one another at the Straits of Mackinac. The Upper and Lower Peninsulas face each other across five miles of open water with a depth of 295 feet.

Native Americans traversed this region for centuries before the first French Explorer's canoe came onto the scene searching for a route to China.

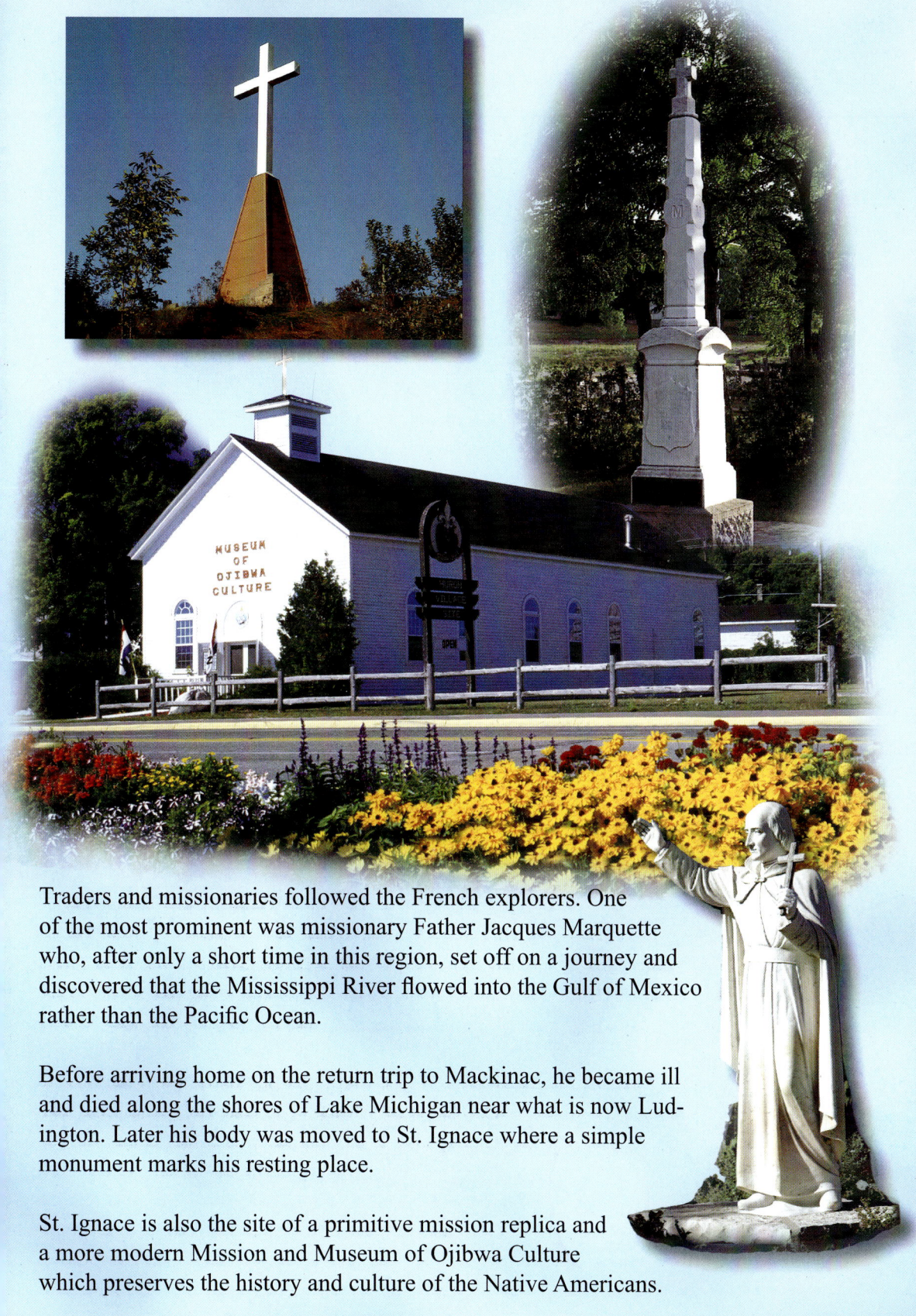

Traders and missionaries followed the French explorers. One of the most prominent was missionary Father Jacques Marquette who, after only a short time in this region, set off on a journey and discovered that the Mississippi River flowed into the Gulf of Mexico rather than the Pacific Ocean.

Before arriving home on the return trip to Mackinac, he became ill and died along the shores of Lake Michigan near what is now Ludington. Later his body was moved to St. Ignace where a simple monument marks his resting place.

St. Ignace is also the site of a primitive mission replica and a more modern Mission and Museum of Ojibwa Culture which preserves the history and culture of the Native Americans.

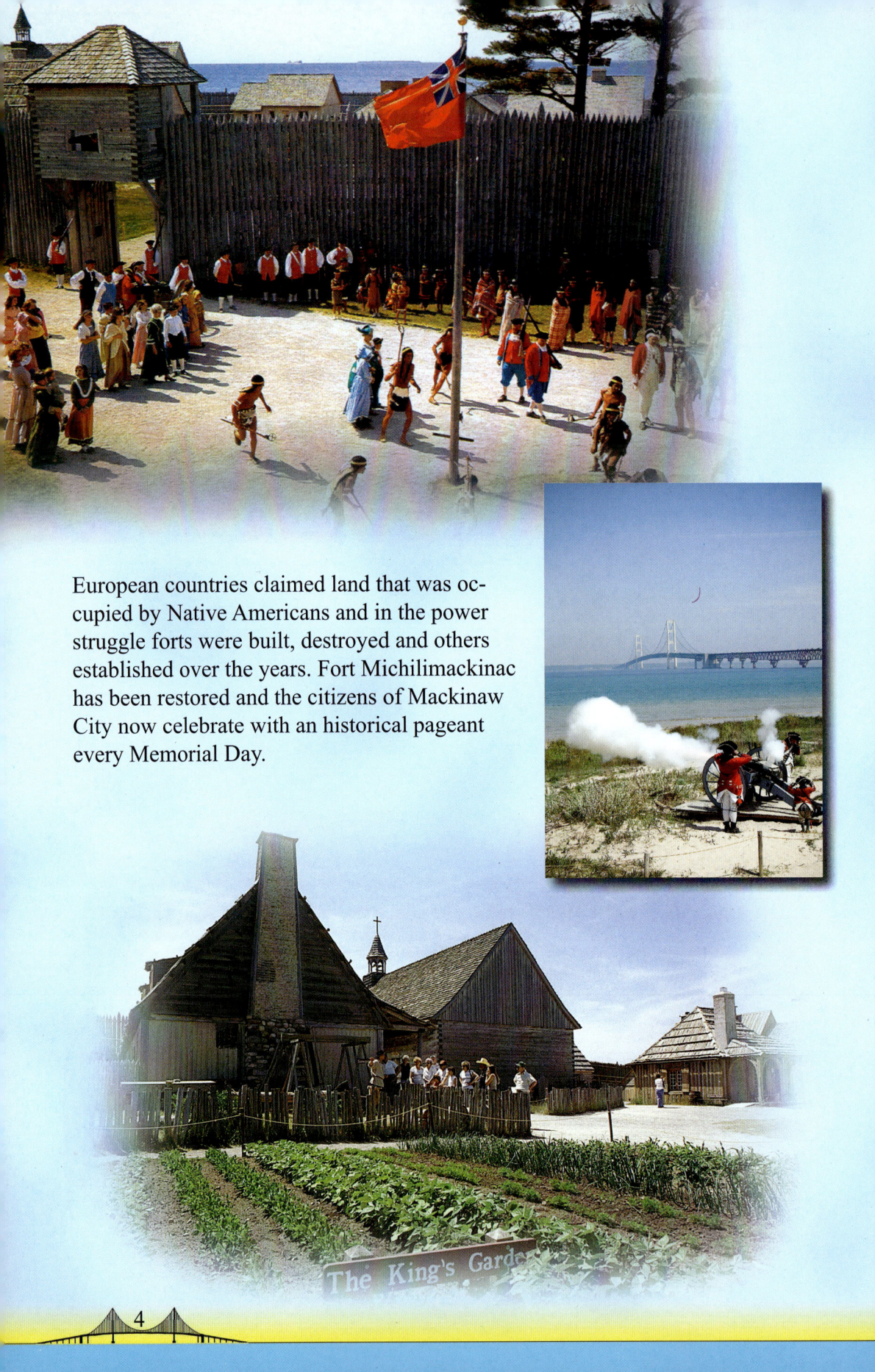

European countries claimed land that was occupied by Native Americans and in the power struggle forts were built, destroyed and others established over the years. Fort Michilimackinac has been restored and the citizens of Mackinaw City now celebrate with an historical pageant every Memorial Day.

Replica of the *Welcome*

In the early days travel was mainly east – west. The first sailing ship in the area was the Griffin which sailed through heading westward. The ship perished on the return trip.

Another sailing ship built in the Great Lakes was the *Welcome*. Shipping soon became commonplace and Mackinac Island became the primary trading center. The rich heritage of the region can be seen everywhere. In St. Ignace the rudder of the wooden steamer *William H. Barnum* is on display along with the windlass from the ship *J. F. Johnson* which sank in 1891.

Sainte Marie

Chief Wawatam

City of Munising

The Straits of Mackinac

Vacationland

Ariel

As the population increased so did the need for transportation between Michigan's two peninsulas. In 1882 the State of Michigan started railroad, mail and passenger ferry boat service between Mackinaw City and St. Ignace. Auto ferry service started in 1923 and lasted 34 years until the Mackinac Bridge was opened in 1957.

City of Munising

Vacationland

The Straits of Mackinac

Special ferry boats also transported railroad cars across the Straits and during the winter often served as ice breakers to keep the channel open for auto ferries as well as iron ore carriers.

U.S. Coast Guard Cutter Mackinaw

For many years there were dreams of a bridge that could cross the impossible Straits. The common belief was that the distance was too far and the water too deep. Adventuresome leaders probed the possibilities and finally the dream became a reality. Dr. D.B. Steinman was selected as designer and architect for the project and soon construction was underway.

Every detail of the bridge building project was planned with amazing precision and infinite detail from the location of the crossing to the height of the towers and the length of the span.

With the foundations firmly in place the aerial part of the construction was next. Iron workers on the bridge took to the sky daily with routines that would equal or surpass any circus high wire act imaginable

Ribbon Tying Ceremony

The impossible dream became possible and the amazing construction was completed on time. It was a great day for Michigan when the Mackinac Bridge opened for traffic on November 1, 1957. Formal dedication ceremonies were held in June, 1958 when Michigan's governor, G. Mennen Williams, Bridge Authority Chairman, Prentiss M. Brown, and their wives did not cut a ribbon but tied together a 22,700 foot ribbon stretching from shore to shore symbolizing the unity of Michigan.

In recognition of this historic event, the U.S. Post Office issued a first class three cent commemorative stamp.

With the opening of the Mackinac Bridge one era was closed and a new page of progress in Michigan's history was turned. It was the end of the automobile ferry boat service between Mackinaw City and St. Ignace and the Old Mackinac Point Lighthouse beacon was extinguished forever. Navigation lights on the new bridge now guided the ship captains. The lighthouse started service in 1892 and served the Straits of Mackinac for 65 years. This beautiful lighthouse landmark, at the foot of the bridge, is now restored to circa 1910 and is open for tours.

Before the bridge was built winter brought great difficulty for the ferry boats when ice frequently blocked the passageway. Now the bridge dominates the stark winter landscape and cars pass over the icy straits without hesitation.

Brilliant shafts of lightning illuminate the sky during a summer thunderstorm.

Mackinaw City offers many bridge viewing possibilities and locations. One is Fort Michilimackinac which stands in bold contrast to the Mackinac Bridge not only in construction materials but also in the span of time between the founding of the fort and the building of the bridge.

The land elevation at St. Ignace is much higher than the Straits of Mackinac so Highway U.S. 2 provides several scenic turn-outs for bridge viewing. On the Upper Peninsula side are the Mackinac Bridge Authority offices and a lovely park and Bridge Welcome Center. It is worth spending a few minutes looking at the pictures and listening to the narration of the bridge history.

Night offers challenges for any photographer attempting to capture images of the bridge. The bridge has flood lights illuminating the towers as well as lights along the sweeping cables which together can make spectacular shots. Pictures taken at twilight will have some subdued color remaining in the sky. Later the sky will be black. The towers are a long distance from shore so a telephoto lens or setting is a must. When it is calm, beautiful shimmering reflections are present in the water. Rough water will produce little or no reflections. A tripod will hold the camera steady as night exposures take much longer than one might think. The Bridge and Lightning picture on pages 16 – 17 was exposed for six minutes; however that was to allow time for more lightning flashes to record.

*The light gleams on my strands and bars
In glory when the sun goes down.
I spread a net to hold the stars
And wear the sunset as my crown.*

D.B. Steinman

An up close and personal view of the Mackinac Bridge is available to anyone who wishes to show up in St. Ignace on Labor Day morning for the annual Bridge Walk. Many thousands of people enjoy this fun-filled festive event. It is a five mile walk across the bridge to Mackinaw City.

Walking along one can feel the strength of the bridge underfoot and watch the huge cables sweep down to eye level at center span, then rise high in the sky as they loop over the towers and once more return to eye level. The experience of being on such a massive structure makes one feel very small and insignificant and at the same time inspires a profound respect and admiration for the designers and builders of the bridge.

When the bridge was built it was said it would last for a hundred years but it is so well designed that some workers said it would last a thousand years. With tender loving care it is in perfect condition now a half century later. The maintenance duties seem endless as some things get replaced from normal wear like roadway gratings and light bulbs. Of course the light bulbs are not easy to reach! There are constant inspections as shown in one picture where the main cable is uncovered and wedges are driven between the cable strands for a peek inside to be sure there is no deterioration. A painting crew is always kept busy. To blend into the landscape the horizontal sections of the bridge are painted green and the vertical towers are ivory.

It takes a crew of over 100 employees, plus some outside contractors to operate and maintain the bridge. There is a security boat always ready at the St. Ignace side and a boat is located below the bridge whenever workers are present above.

Design And Detail Drawings

Total Number of Engineering Drawings 4,000
Total Number of Blueprints 85,000

Rivets And Bolts

Total Number of Steel Rivets 4,851,700
Total Number of Steel Bolts 1,016,600

Men Employed

Total at the Bridge site 3,500
At Quarries, Shops, Mills. etc. 7,500
Total Number of Engineers 350

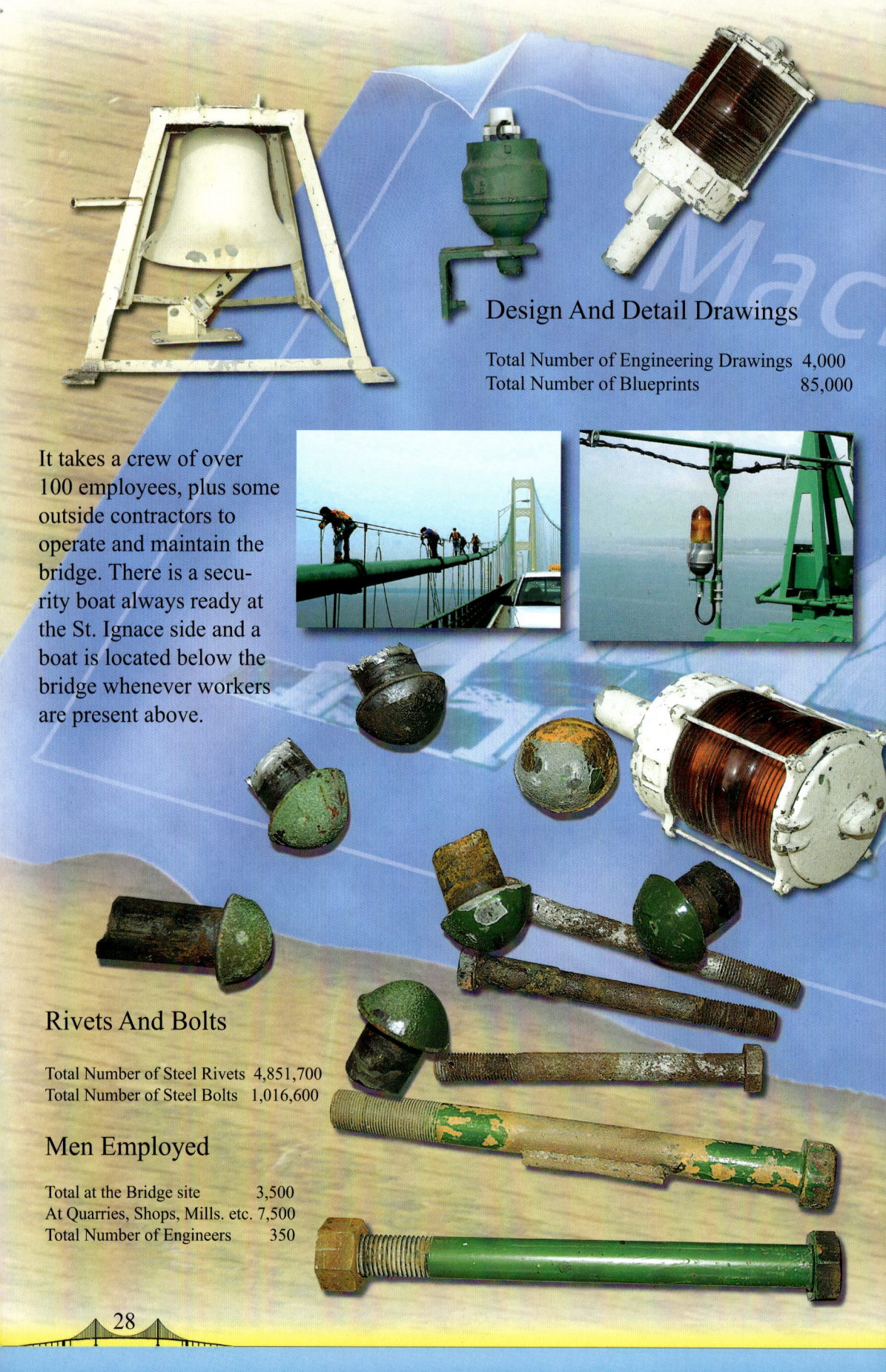

Facts & Figures

The Mackinac Bridge is currently the third longest suspension bridge in the world. The bridge opened for traffic on November 1, 1957

Lengths

Total Length of Bridge (5 Miles)	26,372 Ft.	8,038 Meters
Total Length of Steel Superstructure	19,243 Ft.	5,865 Meters
Length of Suspension Bridge (including Anchorages)	8,614 Ft.	2,626 Meters
Total Length of North Approach	7,129 Ft.	2,173 Meters
Length of Main Span (between Main Towers)	3,800 Ft.	1,158 Meters

Heights And Depths

Height of Main Towers above Water	552 Ft.	168.25 Meters
Maximum Depth of Water at Midspan	295 Ft.	90 Meters
Height of Roadway above Water at Midspan	199 Ft.	61 Meters
Under-clearance at Midspan for Ships	155 Ft.	47 Meters
Maximum Depth of Water at Piers	142 Ft.	43 Meters
Maximum Depth of Piers Sunk through Overburden	105 Ft.	32 Meters

Cables

Total Length of Wire in Main Cables	42,000 Miles	67,592 km
Maximum Tension in Each Cable	16,000 Tons	14,515,995 kg
Number of Wires in Each Cable	12,580	
Weight of Cables	11,840 Tons	10,741,067 kg
Diameter of Main Cables	24 1/2 inches	62.23 cm
Diameter of Each Wire	0.196 inches	.498 cm

Weights

Total Weight of Bridge	1,024,500 Tons	929,410,766 kg
Total Weight of Concrete	931,000 Tons	844,589 kg
Total Weight of Substructure	919,100 Tons	833,793,495 kg
Total Weight of Two Anchorages	360,380 Tons	326,931,237 kg
Total Weight of Two Main Piers	318,000 Tons	288,484,747 kg
Total Weight of Superstructure	104,400 Tons	94,710,087 kg
Total Weight of Structural Steel	71,300 Tons	64,682,272 kg
Total Weight of Cable Wire	11,840 Tons	10,741,067 kg

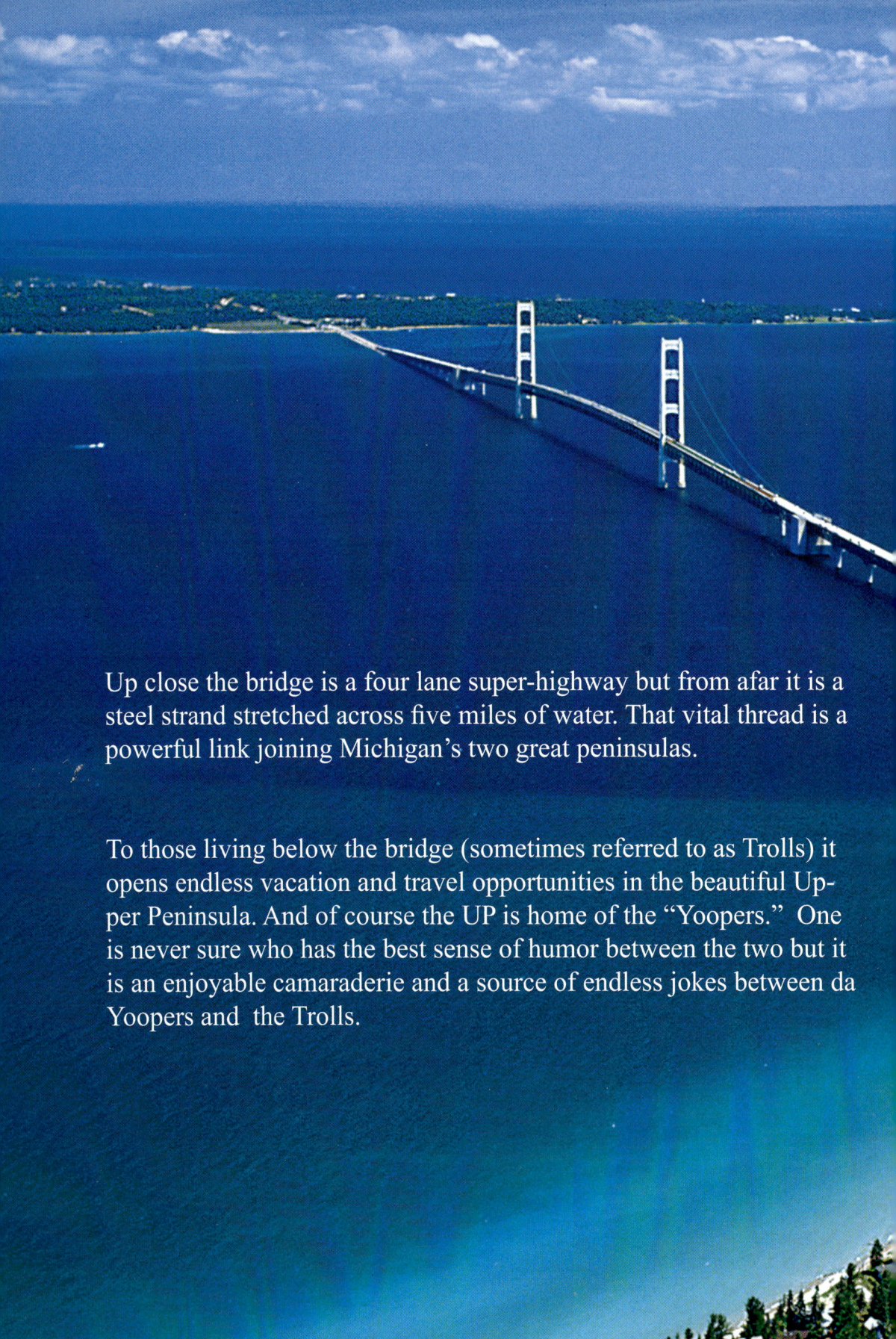

Up close the bridge is a four lane super-highway but from afar it is a steel strand stretched across five miles of water. That vital thread is a powerful link joining Michigan's two great peninsulas.

To those living below the bridge (sometimes referred to as Trolls) it opens endless vacation and travel opportunities in the beautiful Upper Peninsula. And of course the UP is home of the "Yoopers." One is never sure who has the best sense of humor between the two but it is an enjoyable camaraderie and a source of endless jokes between da Yoopers and the Trolls.

Important Dates

Mackinac Bridge Authority Appointed	June, 1950
Board of Three Engineers Retained	June, 1950
Report of Board of Engineers	January, 1951
Financing and Construction Authorized by Legislature	April 30, 1952
D.B. Steinman Selected as Engineer	January, 1953
Preliminary Plans and Estimates Completed	March, 1953
Construction Contracts Negotiated	March, 1953
Bids Received for Sale of Bonds	December 17, 1953
Began Construction	May 7, 1954
Open to Traffic	November 1, 1957
Formal Dedication	June 25-28, 1958
50 Millionth Crossing	September 25, 1984
40th Anniversary Celebration	November 1, 1997
100 Millionth Crossing	June 25, 1998
50th Anniversary Celebration	November 1, 2007